YOUR KNOWLEDGE HAS VALUE

Evidence-based studies on Mycotoxin Levels in Cereals

Awung Elvis

Bibliographic information published by the German National Library:

The German National Library lists this publication in the National Bibliography; detailed bibliographic data are available on the Internet at http://dnb.dnb.de.

ISBN: 9783346921161
This book is also available as an ebook.

© GRIN Publishing GmbH
Trappentreustraße 1
80339 München

Print and binding: Books on Demand GmbH, Norderstedt, Germany
Printed on acid-free paper from responsible sources.

The present work has been carefully prepared. Nevertheless, authors and publishers do not incur liability for the correctness of information, notes, links and advice as well as any printing errors.

GRIN web shop: https://www.grin.com/document/1380130

Evidence-based studies on mycotoxin levels in cereals

By

Awung Nkeze Elvis

ABSTRACT

Organic cereals have lower mycotoxin levels compared to conventionally grown cereals, with organic oats and barley having lower levels. Environmental factors and agronomic practices also affect mycotoxin contamination. Canadian breakfast cereals contain multiple mycotoxins, including deoxynivalenol, fumonisins, ochratoxin A, and zearalenone. Studies aim to understand mycotoxin levels, co-occurrence, and mitigation strategies to enhance food safety.

Mycotoxin contamination in cereals can result from various factors, including the natural co-occurrence of mycotoxins, fungi, animal diets, agronomic and climatic factors, and processing. Fermentation of dough can reduce OTA levels, while climate change impacts mycotoxin levels. Organic cereals have lower Fusarium infestations and contamination. Strategies include biological control agents, bioactive plant metabolites, raw cereal quality, breeding tolerant genotypes, and monitoring contamination. Improve cereal food safety by understanding mycotoxin levels, co-occurrence, and mitigation strategies. Good agricultural practices, plant disease management, and storage conditions limit mycotoxin levels. Food processing techniques, detection technologies, and resistant maize genotypes can reduce mycotoxin content and nutritive value.

Understanding mycotoxin levels, co-occurrence, and mitigation strategies in cereals is crucial for food safety and quality. Mitigation strategies include good agricultural practices, storage and handling techniques, and biological control agents.

Keywords: "studies", "on", " mycotoxin", "levels", "in", "cereals"

INTRODUCTION

Studies on mycotoxin levels in cereals have shown that organic cereals have lower levels of Fusarium infestation and mycotoxin contamination compared to conventionally grown cereals (Bernhoft et al., 2010). The levels of mycotoxins, such as HT-2 toxin and T-2 toxin, were significantly lower in organic oats and barley, while deoxynivalenol (DON) and moniliformin (MON) were found in lower concentrations in organically produced wheat (Bernhoft et al., 2010). Environmental factors and agronomic practices have been found to affect mycotoxin contamination in cereals, highlighting the importance of considering these factors in managing mycotoxin levels (Ferrigo et al., 2016). Additionally, the presence of mycotoxins in cereals, such as DON, has been associated with weather conditions during specific growth stages, such as flowering and close to harvest (Hjelkrem et al., 2016). The regular occurrence of multiple mycotoxins, including deoxynivalenol, fumonisins, ochratoxin A, and zearalenone, has been observed in breakfast cereals in the Canadian market (Roscoe et al., 2008). Overall, mycotoxin contamination in cereals poses a significant risk to human and animal health, and efforts are being made to develop accurate methods for detection and quantification of mycotoxins as well as strategies to reduce their levels in cereals (Shanakhat et al., 2018). The general objective of studies on mycotoxin levels in cereals is to understand the occurrence, variation, and potential health risks associated with mycotoxin contamination in cereals. Specific objectives include investigating the levels of specific mycotoxins, such as ochratoxin A (OTA), deoxynivalenol (DON), 3-acetyldeoxynivalenol (3-ADON), nivalenol (NIV), aflatoxins (AFs), zearalenone (ZEA), fumonisins (FUM), and trichothecenes (TCTs), in cereals and cereal products (Valle-Algarra et al., 2009; Smith et al., 2016). Additionally, studies aim to identify and quantify the co-occurrence of multiple mycotoxins in cereals (Smith et al., 2016; Colli et al., 2021; Zhang et al., 2018). The research also focuses on developing accurate methods for mycotoxin analysis in cereals (Shanakhat et al., 2018; Varga et al., 2012). Furthermore, investigations explore strategies to reduce mycotoxin contamination in cereals through standard processing methods or specific procedures (Shanakhat et al., 2018). Studies have shown that organic cereals have lower levels of Fusarium infestation and mycotoxin contamination compared to conventional cereals (Bernhoft et al., 2010). The presence and amounts of mycotoxins in infant cereals and breakfast cereals are also surveyed (Zhang et al., 2018; Al-Taher et al., 2017). Overall, the objective is to enhance food safety by understanding mycotoxin levels, co-occurrence, and potential mitigation strategies in cereals.

DISCUSSION AND IMPLICATIONS

The occurrence, variation, and potential health risks associated with mycotoxin contamination in cereals can be understood through several factors. Firstly, most fungi are capable of producing multiple mycotoxins simultaneously, leading to the natural co-occurrence of mycotoxins in cereals (Smith et al., 2016). Additionally, cereals can be contaminated by multiple fungi simultaneously or in quick succession, further contributing to the variation in mycotoxin contamination (Smith et al., 2016). Furthermore, animal diets, particularly those of ruminants, often consist of multiple grain sources, increasing the risk of multi-contamination exposure (Smith et al., 2016). Agronomic and climatic factors also play a role in mycotoxin contamination, with studies showing that crop rotation with non-cereals and the absence of mineral fertilizers and herbicides can significantly reduce Fusarium infestation and mycotoxin concentrations in cereals (Bernhoft et al., 2012). Finally, the processing of cereals, such as fermentation and baking, can lead to changes in mycotoxin levels, with some mycotoxins being reduced during fermentation (Valle-Algarra et al., 2009). Overall, understanding these factors can help in developing strategies to minimize mycotoxin contamination in cereals and mitigate potential health risks (Gonçalves et al., 2019).

Mycotoxins, including ochratoxin A (OTA), deoxynivalenol (DON), 3-acetyldeoxynivalenol (3-ADON), nivalenol (NIV), aflatoxins (AFs), zearalenone (ZEA), fumonisins (FUM), and trichothecenes (TCTs), are frequently found in cereals and cereal products (Valle-Algarra et al., 2009). The levels of these mycotoxins can vary during the bread-making process, with fermentation of the dough leading to a significant reduction in OTA levels (Valle-Algarra et al., 2009). Climate change can also impact mycotoxin levels, with projections suggesting that mycotoxin levels associated with cereal diseases may exceed EU limits by 2050 (Vermeulen et al., 2012). Organic cereals have been found to have lower levels of Fusarium infestation and mycotoxin contamination compared to conventional cereals (Bernhoft et al., 2010). A study on cereal-based baby foods found that aflatoxins and DON were frequently detected, with some samples exceeding EU maximum levels (Herrera et al., 2019). The natural co-occurrence of multiple mycotoxins, including trichothecenes, has been observed in unprocessed oats grown in Ireland (Colli et al., 2021). Processing wheat for human consumption can affect mycotoxin levels, with some mycotoxins being reduced during processing (Schaarschmidt & Fauhl-Hassek, 2018). Toxigenic fungi, such as Aspergillus spp., are commonly found in cereals, and their mycotoxins, including AFs and OTAs, can be detected in cereal samples (Alkuwari et al., 2022). A study in South Korea found that mycotoxins produced by Fusarium sp. were frequently detected in commercial grains (Hong et al., 2017). Cereal processing can lead to a reduction in some mycotoxins, but not all can be completely removed (Rehagel et al., 2022). The contamination of agricultural products with mycotoxins in Latin America has significant impacts on human and animal health as well as the economy (Chulze et al., 2021). Recent studies have shown that multiple mycotoxins frequently co-occur in cereals (Bracarense et al., 2011; Chilaka et al., 2016; Smith et al., 2016; Garon et al., 2006; Lombaert et al., 2003; Roscoe et al., 2008). The toxicity of these mycotoxin combinations cannot always be predicted based on their individual toxicities, as their interactions can be antagonistic, additive, or synergistic (Bracarense et al., 2011). To identify and quantify the co-occurrence of multiple mycotoxins in cereals, high-performance liquid chromatography with fluorescence detection (HPLC-FLD) is commonly used (Alshannaq & Yu, 2017). HPLC-FLD methods have been adapted by international organizations for mycotoxin quantification in cereals (Alshannaq & Yu, 2017). The co-occurrence of specific mycotoxins, such as deoxynivalenol (DON) and zearalenone (ZEN), has been reported in cereals (Vidal et al., 2013). These mycotoxins are produced by Fusarium fungi, which are capable of producing multiple mycotoxins concurrently (Smith et al., 2016; Vidal et al., 2013). The regular occurrence of multiple mycotoxins in cereals has also been observed in studies on cereal-based infant foods and breakfast cereals (Lombaert et al., 2003; Roscoe et al., 2008). These findings highlight the need for methods to analyze cereals for multiple mycotoxins (Garon et al., 2006). LC-MS/MS methods are commonly used for mycotoxin quantification due to their ability to detect multiple mycotoxins and their conjugated forms (Nathanail et al., 2015).

To reduce mycotoxin contamination in cereals, several strategies can be implemented. One approach is the use of biological control agents (BCAs) and bioactive plant metabolites, which can help reduce the use of fungicides and control the infection of pathogenic fungi on cereal crops (Ferrigo et al., 2016). Additionally, the quality of the raw cereal used in processing plays a crucial role in mycotoxin reduction. Good-quality cereal grains should be used to minimize mycotoxin concentrations during processing (Chilaka et al., 2016). Breeding mycotoxin-tolerant genotypes is another promising strategy for reducing mycotoxin contamination in cereals (Atanasova-Penichon et al., 2016). Implementing good practices at the preharvest, harvest, and postharvest stages of the crop chain is essential to prevent fungal growth and mycotoxin production (Gonçalves et al., 2019). The danger of mycotoxins in cereals can also be significantly decreased by monitoring the contamination of grains with mycotoxins, improving our understanding of how mycotoxins are distributed during processing, and considering alternate production methods (Pinotti et al., 2016). All things considered, a mix of these techniques can help reduce mycotoxin contamination in cereals and guarantee food safety.

Numerous studies have shown that as compared to conventional cereals, organic cereals typically have lower levels of mycotoxin contamination and Fusarium infection. For instance, a study indicated that when compared to conventional cereals, organic cereals had much lower levels of mycotoxins like deoxynivalenol (DON) and zearalenone (ZEA) and Fusarium species. Mycotoxin and Fusarium infection levels were found to be lower in organic cereals according to similar findings from another investigation. According to these findings, Fusarium infection and mycotoxin contamination in cereals may be decreased by using natural pesticides and fertilizers in organic agricultural practices. Additional research is required to fully understand the mechanisms driving these variances and to validate these findings on a larger scale.

The presence and amounts of mycotoxins in infant cereals and breakfast cereals have been surveyed in several studies. Gratz et al. (2013) conducted in vitro experiments to assess the activity of the human fecal microbiota in releasing deoxynivalenol (DON) from DON-3--D-glucoside (D3G), a plant metabolite of DON. They found that the human microbiota can release DON from D3G and also degrade DON to a less toxic metabolite. Roscoe et al. (2008) conducted a 3-year survey of mycotoxins in breakfast cereals from the Canadian retail market and found that deoxynivalenol was the most frequently detected mycotoxin. Al-Taher et al. (2017) detected DON in both infant cereals and breakfast cereals in the US market. Zhang et al. (2018) conducted a mycotoxin survey of infant and toddler foods and breakfast cereals in the United States and found that deoxynivalenol was the most frequently detected mycotoxin. Overall, these studies demonstrate the regular occurrence of mycotoxins, including deoxynivalenol, in infant cereals and breakfast cereals.

In conclusion, research on mycotoxin concentrations, co-occurrence, and potential mitigation strategies in cereals is crucial if we want to increase food safety and safeguard the caliber and safety of our food supply. Mycotoxins are harmful compounds produced by fungi that can gravely jeopardize human health by infecting crops. In order to create effective mitigation strategies, researchers can learn vital information about the prevalence and distribution of mycotoxin levels in grains. By using mitigation strategies such proper farming practices, appropriate handling and storage techniques, and the use of biological control agents, mycotoxin contamination in grains can be reduced. The outcomes of this study will have an impact on both the production of nourishing food and the preservation of the health of the general people.

CONCLUSION

In conclusion, the objective of enhancing food safety by understanding mycotoxin levels, co-occurrence, and potential mitigation strategies in cereals is crucial for ensuring the quality and safety of our food supply. Mycotoxins are toxic compounds produced by fungi that can contaminate cereals and pose a significant risk to human health. By studying mycotoxin levels and their co-occurrence in cereals, researchers can gain valuable insights into the prevalence and distribution of these toxins, which can inform effective mitigation strategies. Implementing mitigation strategies such as good agricultural practices, proper storage and handling techniques, and the use of biological control agents can help reduce mycotoxin contamination in cereals. This research is essential for safeguarding public health and ensuring the production of safe and high-quality food.

REFERENCES

Alkuwari, A., Hassan, Z. U., Zeidan, R., Al-Thani, R., & Jaoua, S. (2022, June 13). Occurrence of Mycotoxins and Toxigenic Fungi in Cereals and Application of Yeast Volatiles for Their Biological Control. *Toxins*, *14*(6), 404. https://doi.org/10.3390/toxins14060404

Alshannaq, A., & Yu, J. H. (2017, June 13). Occurrence, Toxicity, and Analysis of Major Mycotoxins in Food. *International Journal of Environmental Research and Public Health*, *14*(6), 632. https://doi.org/10.3390/ijerph14060632

Al-Taher, F., Cappozzo, J., Zweigenbaum, J., Lee, H. J., Jackson, L., & Ryu, D. (2017, February). Detection and quantitation of mycotoxins in infant cereals in the U.S. market by LC-MS/MS using a stable isotope dilution assay. *Food Control*, *72*, 27–35. https://doi.org/10.1016/j.foodcont.2016.07.027

Atanasova-Penichon, V., Barreau, C., & Richard-Forget, F. (2016, April 22). Antioxidant Secondary Metabolites in Cereals: Potential Involvement in Resistance to Fusarium and Mycotoxin Accumulation. *Frontiers in Microbiology*, *7*. https://doi.org/10.3389/fmicb.2016.00566

Bernhoft, A., Clasen, P. E., Kristoffersen, A., & Torp, M. (2010, June). LessFusariuminfestation and mycotoxin contamination in organic than in conventional cereals. *Food Additives & Contaminants: Part A*, *27*(6), 842–852. https://doi.org/10.1080/19440041003645761

Bernhoft, A., Torp, M., Clasen, P. E., Løes, A. K., & Kristoffersen, A. (2012, July). Influence of agronomic and climatic factors onFusariuminfestation and mycotoxin contamination of cereals in Norway. *Food Additives & Contaminants: Part A*, *29*(7), 1129–1140. https://doi.org/10.1080/19440049.2012.672476

Bracarense, A. P. F. L., Lucioli, J., Grenier, B., Drociunas Pacheco, G., Moll, W. D., Schatzmayr, G., & Oswald, I. P. (2011, September 22). Chronic ingestion of deoxynivalenol and fumonisin, alone or in interaction, induces morphological and immunological changes in the intestine of piglets. *British Journal of Nutrition*, *107*(12), 1776–1786. https://doi.org/10.1017/s0007114511004946

Chilaka, C., De Boevre, M., Atanda, O., & De Saeger, S. (2016, November 18). Occurrence of Fusarium Mycotoxins in Cereal Crops and Processed Products (Ogi) from Nigeria. *Toxins*, *8*(11), 342. https://doi.org/10.3390/toxins8110342

Chulze, S., Torres, A., Torres, O., Mallmann, C. (2021). Foreword – Special Issue Mycotoxins In Latin America. World Mycotoxin Journal, 3(14), 241-245. https://doi.org/10.3920/wmj2021.x003

Colli, L., Ruyck, K., Abdallah, M., Finnan, J., Mullins, E., Kildea, S., … & Danaher, M. (2021). Natural Co-occurrence Of Multiple Mycotoxins In Unprocessed Oats Grown In Ireland With Various Production Systems. Toxins, 3(13), 188. https://doi.org/10.3390/toxins13030188

De Colli, L., De Ruyck, K., Abdallah, M. F., Finnan, J., Mullins, E., Kildea, S., Spink, J., Elliott, C., & Danaher, M. (2021, March 4). Natural Co-Occurrence of Multiple Mycotoxins in Unprocessed Oats Grown in Ireland with Various Production Systems. *Toxins*, *13*(3), 188. https://doi.org/10.3390/toxins13030188

Ferrigo, D., Raiola, A., & Causin, R. (2016, May 13). Fusarium Toxins in Cereals: Occurrence, Legislation, Factors Promoting the Appearance and Their Management. *Molecules, 21*(5), 627. https://doi.org/10.3390/molecules21050627

Garon, D., Richard, E., Sage, L., Bouchart, V., Pottier, D., & Lebailly, P. (2006, March 31). Mycoflora and Multimycotoxin Detection in Corn Silage: Experimental Study. *Journal of Agricultural and Food Chemistry, 54*(9), 3479–3484. https://doi.org/10.1021/jf060179i

Gonçalves, A., Gkrillas, A., Dorne, J. L., Dall'Asta, C., Palumbo, R., Lima, N., Battilani, P., Venâncio, A., & Giorni, P. (2019, February 12). Pre- and Postharvest Strategies to Minimize Mycotoxin Contamination in the Rice Food Chain. *Comprehensive Reviews in Food Science and Food Safety, 18*(2), 441–454. https://doi.org/10.1111/1541-4337.12420

Gratz, S. W., Duncan, G., & Richardson, A. J. (2013, March 15). The Human Fecal Microbiota Metabolizes Deoxynivalenol and Deoxynivalenol-3-Glucoside and May Be Responsible for Urinary Deepoxy-Deoxynivalenol. *Applied and Environmental Microbiology, 79*(6), 1821–1825. https://doi.org/10.1128/aem.02987-12

Herrera, M., Bervis, N., Carramiñana, J., Juan, T., Herrera, A., Ariño, A., ... & Lorán, S. (2019). Occurrence and Exposure Assessment Of Aflatoxins And Deoxynivalenol In Cereal-based Baby Foods For Infants. *Toxins, 3*(11), 150. https://doi.org/10.3390/toxins11030150

Hjelkrem, A. G. R., Torp, T., Brodal, G., Aamot, H. U., Strand, E., Nordskog, B., Dill-Macky, R., Edwards, S. G., & Hofgaard, I. S. (2016, December 1). DON content in oat grains in Norway related to weather conditions at different growth stages. *European Journal of Plant Pathology, 148*(3), 577–594. https://doi.org/10.1007/s10658-016-1113-5

Hong, S., Kang, J., Cho, S., Lee, K., An, T., Lee, C., … & Chung, S. (2017). Simultaneous Determination Of Multi-mycotoxins In Cereal Grains Collected From South Korea By Lc/ms/ms. Toxins, 3(9), 106. https://doi.org/10.3390/toxins9030106

Karlovsky, P., Suman, M., Berthiller, F., De Meester, J., Eisenbrand, G., Perrin, I., Oswald, I. P., Speijers, G., Chiodini, A., Recker, T., & Dussort, P. (2016, August 23). Impact of food processing and detoxification treatments on mycotoxin contamination. *Mycotoxin Research*, *32*(4), 179–205. https://doi.org/10.1007/s12550-016-0257-7

Klarić, M., Cvetnić, Z., Pepeljnjak, S., & Kosalec, I. (2009, December 1). Co-occurrence of Aflatoxins, Ochratoxin A, Fumonisins, and Zearalenone in Cereals and Feed, Determined by Competitive Direct Enzyme-Linked Immunosorbent Assay and Thin-Layer Chromatography. *Archives of Industrial Hygiene and Toxicology*, *60*(4), 427–434. https://doi.org/10.2478/10004-1254-60-2009-1975

Lombaert, G. A., Pellaers, P., Roscoe, V., Mankotia, M., Neil, R., & Scott, P. M. (2003, May). Mycotoxins in infant cereal foods from the Canadian retail market. *Food Additives and Contaminants*, *20*(5), 494–504. https://doi.org/10.1080/0265203031000094645

Nathanail, A. V., Syvähuoko, J., Malachová, A., Jestoi, M., Varga, E., Michlmayr, H., Adam, G., Sieviläinen, E., Berthiller, F., & Peltonen, K. (2015, May 3). Simultaneous determination of major type A and B trichothecenes, zearalenone and certain modified metabolites in Finnish cereal grains with a novel liquid chromatography-tandem mass spectrometric method. *Analytical and Bioanalytical Chemistry*, *407*(16), 4745–4755. https://doi.org/10.1007/s00216-015-8676-4

Pinotti, L., Ottoboni, M., Giromini, C., Dell'Orto, V., & Cheli, F. (2016, February 15). Mycotoxin Contamination in the EU Feed Supply Chain: A Focus on Cereal Byproducts. *Toxins*, *8*(2), 45. https://doi.org/10.3390/toxins8020045

Rehagel, C., Akineden, Ö., Usleber, E. (2022). Microbiological and Mycotoxicological Analyses Of Processed Cereal-based Complementary Foods For Infants And Young Children From The German Market. Journal of Food Science, 4(87), 1810-1822. https://doi.org/10.1111/1750-3841.16106

Roscoe, V., Lombaert, G. A., Huzel, V., Neumann, G., Melietio, J., Kitchen, D., Kotello, S., Krakalovich, T., Trelka, R., & Scott, P. M. (2008, March). Mycotoxins in breakfast cereals from the Canadian retail market: A 3-year survey. Food Additives & Contaminants: Part A, 25(3), 347–355. https://doi.org/10.1080/02652030701551826

Schaarschmidt, S., Fauhl-Hassek, C. (2018). The Fate Of Mycotoxins During the Processing Of Wheat For Human Consumption. Comprehensive Reviews in Food Science and Food Safety, 3(17), 556-593. https://doi.org/10.1111/1541-4337.12338

Shanakhat, H., Sorrentino, A., Raiola, A., Romano, A., Masi, P., & Cavella, S. (2018, March 13). Current methods for mycotoxins analysis and innovative strategies for their reduction in cereals: an overview. Journal of the Science of Food and Agriculture, 98(11), 4003–4013. https://doi.org/10.1002/jsfa.8933

Silva, A. S., Brites, C., Pouca, A. V., Barbosa, J., & Freitas, A. (2019, November). UHPLC-ToF-MS method for determination of multi-mycotoxins in maize: Development and validation. Current Research in Food Science, 1, 1–7. https://doi.org/10.1016/j.crfs.2019.07.001

Smith, M. C., Madec, S., Coton, E., & Hymery, N. (2016, March 26). Natural Co-Occurrence of Mycotoxins in Foods and Feeds and Their in vitro Combined Toxicological Effects. Toxins, 8(4), 94. https://doi.org/10.3390/toxins8040094

Valle-Algarra, F., Mateo, E., Medina, A., Mateo, F., Gimeno-Adelantado, J., Jiménez, M. (2009). Changes In Ochratoxin a And Type B Trichothecenes Contained In Wheat Flour During Dough

Fermentation And Bread-baking. Food Additives & Contaminants: Part A, 6(26), 896-906. https://doi.org/10.1080/02652030902788938

Varga, E., Glauner, T., Köppen, R., Mayer, K., Sulyok, M., Schuhmacher, R., Krska, R., & Berthiller, F. (2012, February 2). Stable isotope dilution assay for the accurate determination of mycotoxins in maize by UHPLC-MS/MS. *Analytical and Bioanalytical Chemistry*, *402*(9), 2675–2686. https://doi.org/10.1007/s00216-012-5757-5

Vermeulen, S., Campbell, B., Ingram, J. (2012). Climate Change and Food Systems. Annu. Rev. Environ. Resour., 1(37), 195-222. https://doi.org/10.1146/annurev-environ-020411-130608

Vidal, A., Marín, S., Ramos, A. J., Cano-Sancho, G., & Sanchis, V. (2013, March). Determination of aflatoxins, deoxynivalenol, ochratoxin A and zearalenone in wheat and oat based bran supplements sold in the Spanish market. *Food and Chemical Toxicology*, *53*, 133–138. https://doi.org/10.1016/j.fct.2012.11.020

Wan, J., Chen, B., & Rao, J. (2020, March 4). Occurrence and preventive strategies to control mycotoxins in cereal‐based food. *Comprehensive Reviews in Food Science and Food Safety*, *19*(3), 928–953. https://doi.org/10.1111/1541-4337.12546

Xu, R., Kiarie, E. G., Yiannikouris, A., Sun, L., & Karrow, N. A. (2022, June 8). Nutritional impact of mycotoxins in food animal production and strategies for mitigation. *Journal of Animal Science and Biotechnology*, *13*(1). https://doi.org/10.1186/s40104-022-00714-2

Zhang, K., Flannery, B., Oles, C., Zhang, K., Flannery, B. M., Oles, C. J., & Adeuya, A. (2018, April 23). Mycotoxins in infant/toddler foods and breakfast cereals in the US retail market. *Food Additives & Contaminants: Part B*, *11*(3), 183–190. https://doi.org/10.1080/19393210.2018.1451397